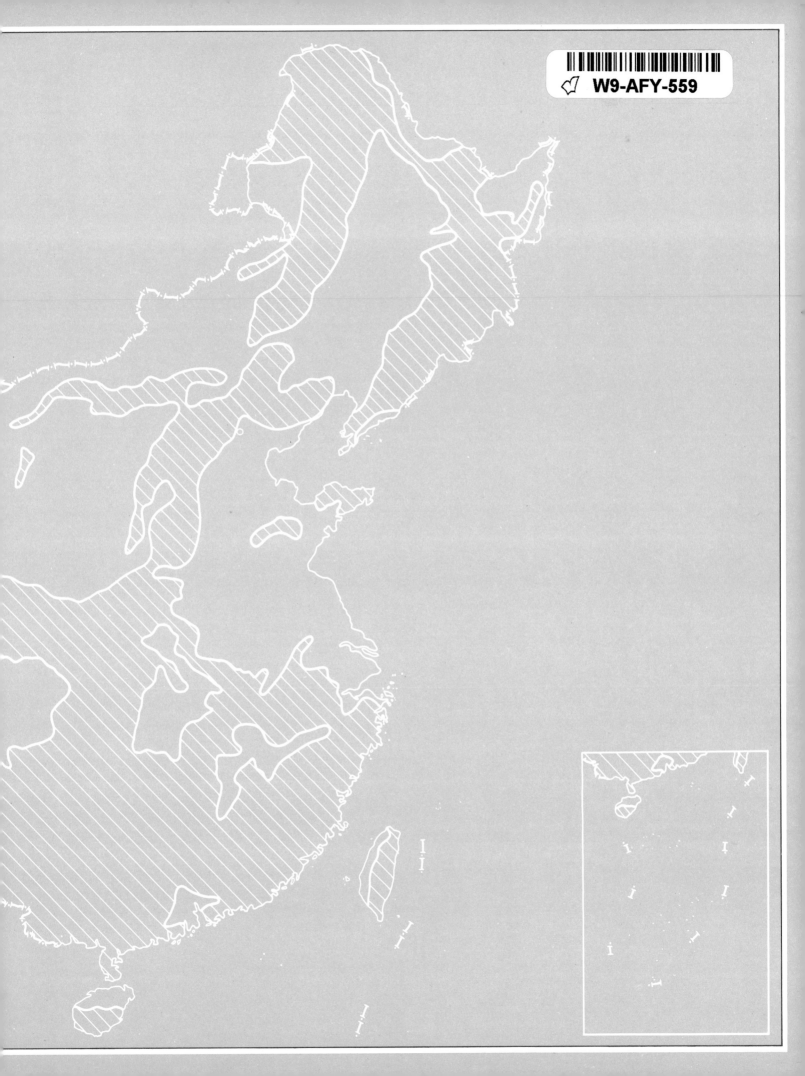

THE ALPINE PLANTS OF CHINA

中國高山植物
THE ALPINE PLANTS OF CHINA

Editor in Chief

Zhang Jingwei (Chang King-wai)

(Division of Biological Sciences, Academia Sinica)

Science Press, Beijing, China, 1982

Gordon and Breach, Science Publishers, Inc., New York

Copyright © 1982 by Science Press and
Gordon and Breach, Science Publishers, Inc.

Published by Science Press, Beijing

Distribution rights throughout the world,
excluding the People's Republic of China,
granted to Gordon and Breach, Science
Publishers, Inc., New York

Printed by C & C Joint Printing Co.,(H.K.)
Ltd.

First published 1982

ISBN 0-677-60190-5

Science Press Book No. 2622-15

Preface

China is a vast country and its territory stretched across three temperature zones: the temperate, subtropical and tropical zones. It is also a mountainous country with 66% of its total area made up of mountains. This great number of mountains provide the necessary diversity of landforms for the creation of a great variety of ecological environments for plants of different forms and natures to grow, multiply, differentiate and survive. Therefore China is one of the countries of the world with the greatest number of and the richest resource of plants and has been known as a "kingdom of plants". According to preliminary statistics, of higher plants alone there are 30000 species of 2988 genera of 300 families. At present about 10000 of the total species have been turned to economic uses.

High mountains in China are concentrated mainly in the southwest of the country. The world's highest, largest and youngest plateau, the Xizang (Tibet) Plateau, was uplifted in this region among high snow-capped and ice-clad mountains. The tallest mountain of the whole world, Mt. Qomolangma (Mt. Everest), is also here, towering on the Sino-Nepalese border, overlooking half of the globe. Because of its extraordinary height, its great number of high mountains and permanent snow, ice and glaciers, the plateau has for ages given people the impression of bleakness, barrenness and extreme cold and windiness. But just lift up a little of this veil of imagination and peep into it, you will be struck by the spectacular panorama of exciting wonderful landscapes and gorgeous plant lives hidden underneath.

Look! On the vast expanse of the flat plateau of 4500 meters a.s.l., the undulating plain formed the chains of low hills and mounds looks like them wavy surface of a heaving sea. Dotted with numerous lakes, pools and swamps. it also looks like a chessboard with the chessmen scattered all over it waging a battle of life and death. These land features also make the plain look like a starstudded night sky.

Below numerous shining snow-capped peaks, on mountain slopes are belts of dense dark green forests and shrubberies. On the broad plain at the foot of mountains, endless lush green grasslands and velvety thick meadows unroll. Oh! How beautiful the Xizang Plateau is! It is another glory of the motherland.

Every year, when spring comes to the plateau and its breeze sweeps gently across the plain, touching every plant it passes by like a magic hand, all the fields and the mountain slopes will be over night covered with flowers of every form of every color and every color shade imaginable to the mind. There are flowers of deep red, brilliant yellow, snow white, blue purple, etc. Blood-red azaleas en masse actually set whole mountain slope aflame.

When summer comes to the plateau, its rainwater and the water from melted snow and ice nourish all the wonderful flowers and rare grasses to great profusion and heavy blossom. It turns the soft green brocade-like grasslands and meadows into countless beautiful multicolored carpets. Glistening ice and snow on mountain tops and dwarfed *Rhodiola* and *Saussurea* wrapped in their white "fur cloaks" exist side by side.

They form a sharp contrast and seem to vie with each other for beauty and glory. The full blown rosy flowers seem to declare proudly to the world that in spite of snow, ice and freezing wind, the plateau is not only with life, but with plenty of life whose beauty, exquisiteness, exotism, and richness in color, form and variety defy imagination.

The alpine plants of the Xizang mountains are flowering plants with beautiful flowers of all forms, sizes and colors. They grow on the mountains like many leis hung around them below snow and ice covered peaks. As soon as our eyes fell on them, we were deeply struck by their unearthly beauty. Wherever we went in the mountains in our expedition, they were always with us. Their beauty, gracefulness and elegance made us forget our fatigue from climbing and inspired us with new courage to face all sorts of hazards and hardships. They were the source of our joy and comfort.

We love the high mountains! But we admire the indomitable spirit of the alpine plants of the plateau and the mountains even more. It is regrettable that these noble plants are greatly attached to their beloved native place. They think nothing of the inclemency of the climate and natural environment of the plateau. They do not want to leave it and no one can make them to. We felt rather sad at the time of parting with the plateau and its magnificent flora, because we were not sure whether we would ever see them again because it is so difficult to visit the place. The only thing we could do was to put what we had seen on celluloid so that we could see them again whenever we want. That is the way the present album was conceived. We published the album so that readers all over the world may see for themselves what a wonderland the Xizang Plateau is and how beautiful the alpine plants there are.

But who are the people who made this album and who make it possible for the reader to share their wonderful experience in the wonderland. Open the album and you will get all the answers you want.

The Xizang Plateau, on account of its geographical position and peculiar natural environment, has a great variety of vegetations with a large number of species of plants. It is even more rich in alpine plants. In higher plants alone, Xizang has 5700 species of 1146 genera of 208 families. If algae, mosses, fungi and lichens are included, the number of species will far surpass 10000. The present album contains about 140 species of higher plants, algae, mosses, fungi and lichens. Through the photographs and the accompanying explanatory notes the readers are taken to the landscapes, ecological environments, and economic and decorative or garden alpine plants of Xizang. As the above figure shows, the photographs of the album represent only 1 — 2 % of the total number of species of the plants of Xizang. It is impossible to see all the plants of the region in the album. Already the album contains more wonderful flowers and plants than the mind can cope with in one reading. But it must be pointed out that the real things, the actual flora of the Xizang Plateau are vastly much more numerous in number and in variety, much more wonderful, strange and beautiful in form and color.

Though small in number, the beautiful color photographs presented in the present album represent the fruits of untold sufferings, hardships and great risks Chinese scientific workers and photographers went through over a dozen years in places and at heights seldom reached by man. Most of these valuable photographs were taken in mountains at heights over 4500m above sea level and some of them heights ranging from 5700 — 6000 m, a.s.l. For instance, the pictures of *Delphinium brunonianum* Royle were taken 6000 m, a.s.l. There are many new species presented in the present album. They are all firsthand materials published for the first time. There is no doubt that they would be invaluable for both scientific research and artistic photography. The album is actually the combination of the two. Its publication will undoubtedly contribute to the development of botany, phytoecology and geobotany. It will serve well as a general reading and reference book for botanists, phytoecologists and geobotanists.

The chief editor of the album is Zhang Jinwei, who has been studying the phytoecology and botany of the Qinghai-Xizang Plateau since 1960 and been to the plateau nine times on scientific expeditions. He is also the editor in chief of the Album of Photographs of the Mt. Qomolangma (Everest) Expedition, published by the Science Press in 1974. He is also the author of the following papers: "The Vegetation of Central Xizang

(Tibet)", "A primary study on the vertical vegetation belt on Mt. Qolmolangma (Everest) region and its relationship with horizonal zone", and "On the vegetation latitudinal zoning of Qinghai-Xizang (Tibet) Plateau".

Editors and photographers of the album also include the following: Zhang Jingwei (Chang King-wai), Wang Jinting, Zhang Hesong, Li Bosheng, and Chen Weilei. Some of the photographs in the album are contributed by Li Wenxiu, Song Kairu, Lang Kaiyong, Bao Xiancheng, Ni Zhicheng, Wei Jiangchun, Wang Shaoqing, Chen Jiayou, Chen Momei, Wu Sugong and Zhao Kuiyi and others.

Editor in Chief

Zhang Jingwei (Chang King-wai)

(Division of Biological Sciences, Academia Sinica)

Compilers

Zhang Jingwei   (Chang King-wai)   Wang Jinting

Zhang Hesong   Li Bosheng   Chen Weilei

Photographers and Photo Contributors

Zhang Jingwei   Wang Jinting   Zhang Hesong

Li Bosheng   Chen Weilei   Li Wenxiu

Song Kairu   Lang Kaiyong   Bao Xiancheng

Ni Zhicheng   Wei Jiangchun   Wang Shaoqing

Chen Jiayou   Chen Momei   Wu Sugong

Zhao Kuiyi

Mount Qomolangma (Everest)

Situated on the border between China and Nepal in the middle part of the Himalayas, it is the highest peak of the world, towering 8848 meters a. s. l. and is now still rising at the rate of 5 mm annually owing to the tectonic movement of the plates of the earth crust. Amid a number of less glamourous and lofty peaks, it forms with the Xizang plateau the world's most magnificent region of plateau and mountains.

The Potala Palace
    First built in the 7th century and rebuilt in 1645 on a hill of a plain in the valley of the Lhasa River, which is a tributary of the Yarlung Zangbo River. It is an imposing palace structure 110 meters high, with golden roofs. In it are enshrined the tombs of the Dalai Lamas of all the past dynasties.

Landscapes in the lower reaches of the Yarlung Zangbo River

The river is originated in the snow mountain Zhimayongdanfu, 6532 meters a. s. l., on the north slope of the Himalayas. From its source in the Jiemayangzong glacier eastward to the border where it flows out of the country over the border, it is 2057 kilometers long. It is in altitude the largest river of the world. In its lower reaches, the climate is damp and hot so the region presents many tropical landscapes with many tropical features such as tree fern (*Sphaeropteris brunoniara* (Hk.) Tryon), tree frog (*Rhacophorus maximus* Guenther), cauliflory, plank-buttress.

Sphaeropteris brunoniara (Hk.) Tryon—They are found in Motuo County at an elevation of 900 meters a. s. l.

Musa rubra Wall ex Kurz.

A Bamboo Forest

Rhacophorus maximus Guenther, Rahcophoridae.

The genus *Rhacophorus* has here many species and they are mainly distributed in South and East Asia. In China, they are mainly found in Yunnan and southern Xizang. There are suckers on their toes and fingers, with which they can attach themselves to the trunks and branches of trees.

These shown here are of tropical species and they flourish on the south side of the Himalayas and in Motuo County, Xizang at 900 meters a. s. l.

Tacca integrifolia Ker-Gawl.

The plant is widely distributed in the rain forests of Southeast Asia. The plant shown in the picture here is found on a stream bank below 1200 meters a. s. l. in Motuo County, Xizang. It is also distributed in Yunnan Province. A perennial herb, it belongs to Genus *Hydrome*.

7

Cauliflory.

In humid tropical countries, the climate is hot and damp, so plants do not need thick barks to protect them from cold and drought. Therefore, buds grow on tree trunks and burgeon there into flowers. Moreover in tropical forests, foliage is too dense and thick and flowers on branches and tender twigs are entirely buried under thick leaves and are difficult for pollencarrier insects to discover and get to. So after long ages of adaptation, this method of spreading pollens by growing flowers on trunks that are easy to see and reach was chosen and perpetuated. Of course, it sounds a bit absurd that flowers should be grown on trunks instead of 1—2 years green twigs, but it is nevertheless a fact out of necessity and quite natural in view of the ecological environment of tropical rain forests.

The picture shown here is the *Ficus semicordata* Buch. -Ham. ex J. E. Smith that bears fruits from flowers on old trunks.

Gynocardia odorata R. Br.

This is also a kind of trunk flowering and fruiting tree. Its seeds contain edible oil.

Tropical montane evergreen broadleaf forests.

Such forests are found on the south slope of the Himalayas in a belt beween 1000-2500 meters a. s. l. They are composed mainly of evergreens of the *Quercus*, the *Castanopsis* and the *Lithocarpus*.

This is the evergreen broadleaf forest found in Motuo County, Xizang at 1200 meters a. s. l.

Epiphyte grown in the tropical montane evergreen broadleaf forests.

Bulbophyllum cariniflorum Rchb. f, is an epiphyte grown on stones in forests in Gyirong, Xizang, at altitude 1200 meters (Upper picture).

Dendrobium moniliforme (L.) Sw. is a kind of epiphyte grown on other trees. It is a Chinese medical herb used as a cure for gastritis (Lower picture).

Forest of *Pinus roxburghii* Sarg. (syn. *P. longifolia* Roxb.)

Pinus roxburghii Sarg., is a kind of coniferous arbor distributed in mountain districts of low elevations. In China forests of this tree are only found in Gyirong, Xizang, below 2300 meters a. s. l.

Landscape in Bomi District

Forest of *Pinus griffithii* M'Clelland
It is found in the valley of the Gyirong River in the southern part of Xizang.

Forest of *Pinus densata* Mast.

This is a kind of pine forest widely distributed in the mountains of Sichuan, Xizang, Yunnan and Qinghai. Some botanists regard the trees as a variety of the *Pinus tabulaeformis* Carr. widely distributed in the temperature zone of North China. This shows at least the close relationship between the two. The *Pinus densata* Mast., is distributed in the mountain belt between 2200—3700 meters a. s. l. in Xizang and forms forests between 2500-3500 meters a. s. l.

The above is the picture of one of the forests of *Pinus densata* Mast. on the banks of the Yigong Zangbo River in the southeastern part of Xizang.

Pinus griffithii M'Clelland

    Pinus griffithii M'Clelland is a plant endemic to the Himalayas. Its distribution reaches in the east Burma and the northwestern part of Yunnan, and in the west along the Himalayas, Pakistan and Afghanistan. In altitude it grows in mountain belts from 1000-3300 meters a. s. l. and forms its pure forests in these belts between 2000-3000 meters a, s, l. Its wood is excellent construction timber. It can be tapped for pine resin which can be refined into turpentine. Its seeds contain 20 % oil, which can be used in industry.

Drosera peltata Smith. var. *lunata* (Buch.-Ham.) Clarke

It is a perennial herb used in China as a cure for various aches, poor blood circulation and extravasated blood and various congestions. It is found in meadows in forests of *Pinus griffithii* M'Clelland on the south slope of the Himalayas at Gyirong at 2500 meters a. s. l. It is also an insectivorous plant. Both the edge and surface of its semilunar leaf secrete a kind of sticky substance that will capture small insects coming into contact with it and has them digested and absorbed as food. This can be seen in the picture above.

Roscoea capitata J. E. Sm.

It is a member of the family Zingiberaceae and grows in damp meadows on the fringes of forests and has blue-purple flowers. It has clusters of fleshy rhizomes like spindles and is used in China as a cure for aches, indigestion, belch and various aches in the abdomen.

Archiclematis alternata (Kitamura et Tamura) Tamura

The plant is a primitive species of the genus *Clematis*. All plants of the genus have opposite leaves, but as a primitive species, this plant has alternate leaves. It is a new species discovered on the south side of the Himalayas and the Japanese botanist Tamura considered it a new genus and a new species and named it *Archiclematis alternata* (Kitamura et Tamura) Tamura. It grows in Gyirong, Xizang, at 2200 meters a. s. l.

Euphorbia royleana Boiss.

It is a kind of drought-resistant meaty plant and is found in the dry riverbeds in the valleys of the Nujiang River, the Lancang River, the Jinsha River and in the Western Himalayas. The above is a photo of the plant in a dry riverbed at 1800 m. a. s. l. near the Resuo Bridge, Gyirong, Southern Xizang.

Forest of *Tsuga dumosa* (D. Don) Eichler

The plant is distributed in the montane coniferous-broadleaf forest belt in Xizang, Yunnan and Sichuan. It is one of the principal trees of such forest. In Xizang it is distributed at altitudes between 2500-3500 meters. The above is the photo of a forest of *Tsuga dumosa* (D. Don) Eichler in Mutuo, Xizang, at 2800 meters a. s. l. where the climate is damp and the stones in the forests are covered densely with bryophytes.

Quercus semicarpifolia Smith

It is distribued mainly in the Himalayas. In the southern and eastern parts of Xizang, it often forms montane coniferous-broadleaf forest with *Tsuga dumosa* (D. Don) Eichler at altitudes between 2500-3800 meters. Its fruits contain starch and can be used as fodder for pigs. Its leaves and seeds are also used medically as a cure for diarrhoea, dysentries, enteritis and asthma and as a general anti-toxin purgative. It bears strong resemblance to *Quercus ilex* L. distributed in the Mediterranean region and western Asia and is therefore often regarded as a remnant plant of the ancient Mediterranean region.

Ganoderma oroflorum (Lloyd) Teng
    It is grown in evergreen broadleaf forest at 2000 meters a. s. l.

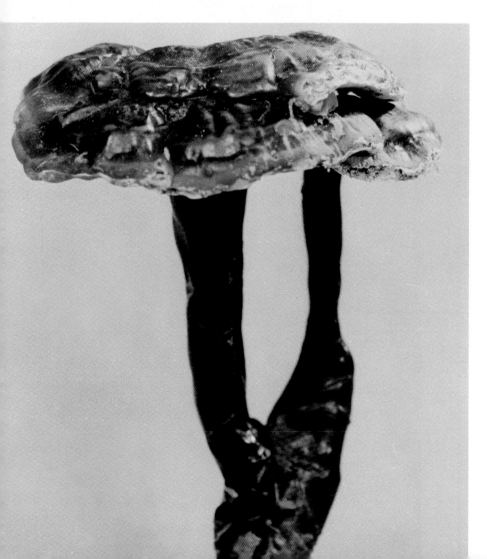

Ganoderma lucidum (Leyss. ex. Fr.) Karst.
    It is grown in the forest of *Quercus semicarpifolia* Smith.

Picea smithiana (Wall.) Boiss.

It is a species of the west Himalayas. In China it is distributed only in Gyirong, Xizang at altitudes between 2300-3200 meters.

Cupressus gigantea Cheng et L. K. Fu

It is a newly discovered cypress at altitudes between 3000—3300 meters in slightly alkaline soil in the lower reaches of the Yarlung Zangbo River and in the valley of one of its trebutaries, the Nyanghe River. It often forms small pure forest. The above is the photo of a giant tree of the plant, about 5.8 meters in circumference and 50 meters tall and its age is estimated at above 2300 years. Its seeds are used in China as a cure for palpitstion and insomnia.

Picea likiangensis (Franch.) Pritz.
var. *balfouriana* (Rehd. et Wils.)
Cheng ex Hu

It is distributed at altitudes between 3200—4100 (4400) meters. In the mountains of Sichuan, Northwest Yunnan and the extensive region of the eastern part of Xizang. It usually forms great thick forest and it is one the excellent timber trees. The above is a photo of the plant in Bomi.

Forest of *Abies spectabilis* (D. Don) Sparch

This is the principal type of subalpine coniferous forest in the southeastern part of Xizang and on the southern side of the Himalayas, at altitudes between 3500—4000 meters.

Abies spectabilis (D. Don) Sparch and its cones

Phellinus pini (Thore ex Fr.) Ames var. *abietis* (Karst.) Bres.

It grows on the living wood of *Picea likiangensis* (Franch.) Pritz. var. *linzhiensis* Cheng et Fu and *Picea likiangensis* (Franch.) Pritz. var. *balfouriana* (Rehd. et Wils.) Cheng ex Hu. It is a representive species of the fungi in the forest of these plants.

Trametes trogii Berk.

It is a parasite plant on the plants of Salix in the valleys of the middle reaches of Yarlung Zangbo River between 3700—4000 meters a. s. l. It is a typical semi-arid alpine-plateau fungus.

00016

Bushes of *Rhododendron wightii* Hk. f.
    Thick bushes of it are found in the valley of the Penqu River and Kadar River in the southern part of Xizang at 4000 meters a. s. l.

Pogogatum sp.
    The mossis grown under subalpine bushes of rhododendron at 4200 meters a. s. l.

The alpine vegetation belt

In mountains, temperature drops as the altitude increases. At altitudes where the average temperature is below 10°C during the growth period of the year, trees can not accumulate enough food for growth, so they are replaced by short shrubs or grass, which are usually distributed above the forest belt and form what is called the alpine vegetation belt. In the above picture, we can see such belt above the dark green forest belt. The demarcation line between the two belts is very clear. On the southern side of the Himalayas, this line runs roughly along a line about 4000 meters a. s. l.

In the alpine vegetation belt are distributed bushes and meadow of all forms and colours, which are gorgeously beautiful and graceful.

Sabina pingii (Cheng) Cheng et W. T. Wang. var. *wilsonii* (Rehd.) Cheng et L. K. Fu bush.

The above picture shows an alpine meadow studded with semicircles of bush of the plant, making it look like a flower bed in an alpine park.

Sabina wallichiana (Hk. f. et Th.) Kom bush

These shrubs are distributed in the southern and eastern parts of Xizang, the northwest part of Yunnan in China, Bhutan and Nepal at 4000-4300 meters a. s. l. and form alpine short bush as shown in the above picture. But in the sub-alpine belt below the upper limit of the forest belt, they grow into subalpine forests of tall coniferous trees as shown in the picture below. This great change of the same plant in different ecological circumstances at different altitudes above sea level shows the great effect of low temperature and strong winds on the growth of plants.

Alpine *Salix* sp.

    Like the *Sabina wallichiana* (Hk. f. et Th.) Kom. this alpine species of *Salix* grows to only 10—20cm at altitudes above 4000 meters but a species of a similar genus, the *Salix paraplesia* Schneid. var. *subintegora* Z. Wang et P. Y. Fu, grows as tall as 10—20 meters at 3670 meters a. s. l. in the Lhasa District.

Rhododendron thomsonii Hk. f. bush
　　This is a picture of the Rh. *thomsonii* bush grown at 3800 meters a. s. l. in the Chentang area, Dinggyê County, Xizang.

Caragana jubata (Pall.) Poir. bush
　　The *Caragana jubata* is a kind of cold-and drought-resistant shrub found on dry gravel slopes in the vicor higher of the subalpine forest belt.

Rhododendron nivale Hk. f. bush

This rhododendron is the most widely distributed high-altitude species in Xizang. This is a picture of a bush of this plant at 4700 meters a. s. l. in the Riwu area, Dinggyê County, Xizang.

Rhododendron setosum D. Don bush

The plant is distributed on the southern side of the Himalayas in the southeastern part of Xizang at altitudes between 3800—4500 meters.

Rhododendron hypenanthum Balf. f. bush
It is one of the alpine shrubs of the Mount Qomolangma area of the Himalayas.
This is a bush of the plant at 4100 meters a. s. l. in Nyalam, Xizang.

Alpine Cushion Vegetation

Cushion vegetation is the plant community formed in the cold and windy environment of high mountains. It is distributed on the northern side of the Himalayas at altitudes ranging from 4800 to 5500 meters. It is composed of *Arenaria musciformis* Edgew. The part above ground is in the form of semicircles spread on gravel slopes. The above is a photograph of such alpine cushion vegetation.

The Ranwu Lake

The lake is situated at the upper source of the Bolong Zangbu River at 3900 meters a. s. l. in the southeastern part of Xizang. It was formed by the blocking up of the river by mud-rock flow. There are forest of *Picea likiangensis* (Franch.) Pritz. var. *balfouriana* (Rehd. et Wils.) Cheng et Hu, on the surrounding mountains. In summer, the lake with its water as serene and clear as a glass forms with the surrounding mountains and forests extremely beautiful landscapes (in the upper photo)

In winter, all forests and mountains around the lake are covered with snow. The great calm and quiet reigning on the lake inspire men with the great composure and imperturbability of nature (in the lower photo).

Evening on the Ranwu Lake

Nostoc fuscescens F. E. Fritsh
This algae is found in the stagnant water of 19° C with a pH value of 6.5 in depressions in the meadows around the Ranwu Lake. *Nostoc fuscescens* F. E. Fritsh was first discovered in 1921 in the South Pole. Its characteristics are: spherical form, drab color, with filaments growing in parallel when young and twisted together when ripe, with cells normally spherical or tubular in form with elliptical variations.

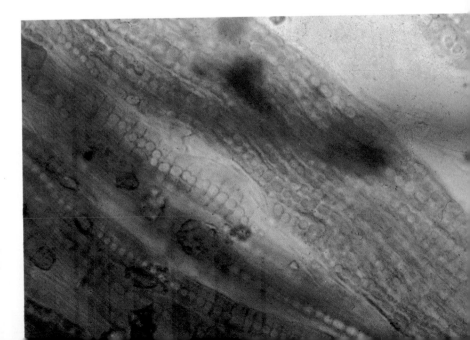

Meadow of *Polygonum viviparum* L.

In damp alpine areas, *Polygonum viviparum* L. grows in meadows with other plants. During the blossom season in meadows where it predominates, it makes them look like plush variegated carpets.

Meadow of *Kobresia pygmaea* C. B. Clarke

The plant is distributed from 4000 to 5200 meters a. s. l. and even higher. Its meadows are excellent local pasture grounds.

A "red lake" at the top of a high mountain

On the southern slope of Eastern Himalayas at Cuona, Xizang, there is a glacial lake formed by the silting up of former glaciers. The surface of the lake is 4200 meters a. s. l. The water in the upper and middle layers of the lake is red in color and forms red belts that run into the lake (the picture below). This is the first "red lake" discovered at high altitudes in mountains in the world. The color of the lake is due to the presence of a great amount of *Glenodinium gymnodinium* Pen. in the water (the middle and the upper right pictures).

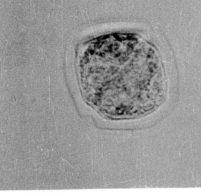

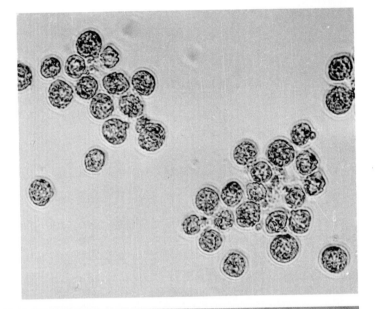

Leptodermis microphylla (H. Winkl.) H. Winkl.

It is distributed in the valley of the Yarlung Zangbo River on arid mountain slopes from 2500—4100 meters a. s. l.

Bush of *Sophora moorcroftiana* (Wall.) Benth. ex Baker

It is distributed on both sides of the upper and middle reaches of the Yarlung Zangbo River on semi-arid mountain slopes and terraces from 3500 to 4200 meters a. s. l. This is a photograph of this bush on mountain slopes near Lhasa, Xizang.

The steppe at the foot of Mount Xixabangma, 8012 meters a. s. l.

Plants in the steppe *Heteropappus boweri*
(Hemsl.) Griers. (syn. *Aster boweri* Hemsl.)
It is found in the Qiangtang plateau in
the northern part of Xizang, at about 5000
meter a. s. l.

44

Thymus linearis Benth

It is distributed in the valley of the Konqi River in the southwestern part of Xizang at about 4000 meters a. s. l. It is one the important alpine steppe plants in the western part of the Central and the Western Himalayas.

Oxytropis glacialis Benth.

Cousinia thomsonii C. B. Clarke

It is a xerophilous plant distributed in the western part of the Central and the Western Himalayas.

The Bangong Lake (Bangong Co)
It lies on the border between China and India in the northwestern part of Xizang at 4241 meters a. s. l. with an area of about 430 square kilometers.

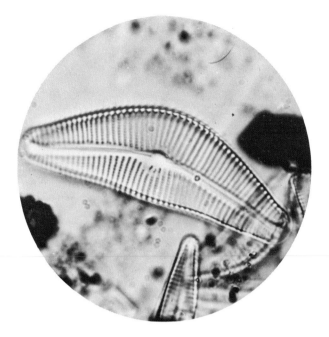

Cymbella cistula (Hemprich) Grun.

It is widely distributed in great quantities in the Transversal mountains, Xizang and Xinjiang It is a plateau and alpine cold water diatom.

Didymosphenia geminata (Lyngbye) M. Schmidt

It is an alpine and plateau alga growing in cold running water. It is widely distributed in the northwestern and southwestern parts of Sichuan province, the northwestern part of Yunnan province, Xizang and the Xinjiang Uygur Autonomous Region. Its distribution in Xizang is characterized by its concentration in the southeastern part of the province and its gradual decrease in amount westward and north-westward in Xizang.

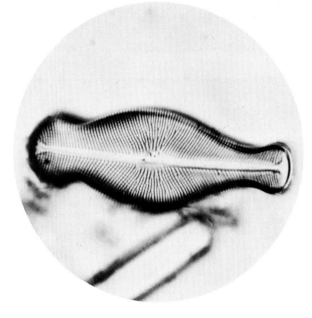

Didymonema tibeticum Wei and Hu——a new species of a new genus.

The present specimen was gathered from the bank of the Zhenquan Lake in North Xizang at 4784 meters a. s. l. It grows in salt lakes of sodium sulfate.

The alga is band-like in form composed of fine long filiform bodies without branches. Except a few cells that are single files at the base of the alga and a part of the cells forming single files of pseudorhize, all the rest cells are arranged in parallel double lines. The vertical section of the "inter-layer" of the horizontal wall of the cell has an H-shaped pattern. There is no other algae of the same family with this structure. So the present specimen is a new species of a new genus of this family.

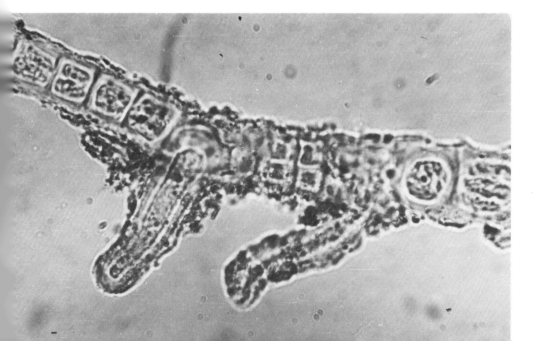

Dracocephalum heterophyllum Benth.
This was found in the Ali district at 4750 meters a. s. l. in western Xizang.

The *Oxytropis microphylla* (Pall.) DC. in the *Stipa purpurea* Griseb. steppe.

Allium przewalskianum Regel.

The plant is distributed in the steppe of northern Xizang. When young, its leaves are tender and delicious and can be eaten as a vegetable. The Shiquan River (Sênggê Zangbo) in the valley of which the plant is found is 4650 meters a. s. l.

Caragana versicolor Benth.

One of the important principal underbrushes in semiarid areas of the Qinghai-Xizang Plateau, the Caragana versicolor Benth. is distributed in Xizang, Qinghai, and West Sichuan on mountain slopes. It blossoms in multitudes of small yellow flowers.

The Yangbajain Hot Spring Lake.

This hot spring lake is about 90 kilometers north of Lhasa at 4300 meters a. s. l. It is formed by a group of hot springs. Geologically the Xizang plateau is rather young and is still in a stage of active tectonic adjustment. It is therefore rich in geothermal resources and hot springs are a ubiquitous sight everywhere.

Algae found around a hot spring.
These are the alga found in the immediate water around a hot spring near Sêwa in north Xizang.

A Hot Spring Near Qamdo
(Changdu)

Phaeonychium parryoides (Kurz ex Hk. f. et Anders) O. E. Scutz.

It is distributed in north Xizang at an altitude of 4600 meters. Covered with white downs, it is a plant found only in Xizang at present.

Cremanthodium plantagineum Maxim.

It is found on the north slope of the Gandisê Range at altitudes between 5400 and 5600 meters. It is used as a medical herb for curing coughing and eliminating expectoration.

Saxifraga nanella Engl. et Irmsch.

One of the common alpine plants in Xizang, it is distributed in north Xizang on mountain slopes between 5200 and 5400 meters a. s. l.

Microula tibetica Benth.

It is found in steppe at 5000 meters a. s. l. in north Xizang.

Desert plant communites of *Ceratoides latens* Rev. et Hol. [syn. *Eurotia ceratoides* (L.) Mey.] *Ajania fruticulosa* Poljark.

The northern part of *Xizang* is thousands of kilometers away from the sea and is completely cut off from the moist air from it by high surrounding mountains. As a result, alpine desert vegetation is developed in arid areas. The above are such plant communities found in Ritu (Rutog) County on mountain slopes at 4530 meters a. s. l.

Ceratoides compacta Tsien et C. G. Ma. (Syn. *Eurotia compacta* A. Los.)

This is an alpine desert plant found up to the present only in the Xizang Plateau. It grows close to the surface of the ground like a cushion, and to adapt itself to the dry, cold and windy ecological environment, its leaves are covered with downs. The above is a photograph of *Ceratoides compacta* Tsien et C. G. Ma. found north of Ritu (Rutog) County, 5050 meters a. s. l.

Christolea crassifolia Cambess
   This is an alpine plant with carneous leaves widely distribued in the steppe and deserts in the north-western part of Xizang. The above is a photo of a community of this plant found in Ritu (Rutog) County at 4650 meters a. s. l.

Physochlaina praealta (Decne.) Miers.

   It is generally distri-buted in Central Asia. It is also found in Burang County, Xizang at places 4600 meters a. s. l.

Ajania fruticulosa Poljark
   It is distributed in Ritu (Rutog) County, Xizang, in places 4530 meters a. s. l. It is an alpine steppe desert plant.

Mount Kangrinboqê

Towering 6656 meters a. s. l., it is the main peak of the Gandisê Range. It is regarded by Buddhists as a sacred mountain and is visited every year by pilgrims from Nepal and India.

On the lakeside of the Mapongyongcuo (Mapam Yumco) Lake

The Mapong yongcuo Lake is situated at the foot of Mount Kangren Boqi at 4588 meters a. s. l. with an area of 410 square meters. It is connected with the Laangcuo Lake, which is 4573 meters a. s. l., 280 square meters in area. It is regarded as a sacred lake by Buddhists, and is visited by Buddhist pilgrims from Nepal and India for bathing in it. It is said a bath in it will free the bother from illness and calamities.

Androsace lhasaensis Yang aff.

Saxifraga pasumensii Merq. et Shaw aff.

Among alpine plants, there are some that grow close to the ground with their leaves spread flat out like the petals of a lotus flower and form a rosette as a result of their adaptation to their ecological environment. This arrangement of their leaves will enable them to have a maximum area for receiving sunlight for photosynthesis. Because of their low-lying position, they are also getting the advantages of cushion plants. This form of growth is found mainly with herbs. *Saxifraga pasumensii* Merq. et Shaw aff. is one of such plants.

57

Rheum spiciforme Royle

The plant has two varieties. The upper picture is the variety that is found at 5000 meters a. s. l. on the top of the Mala mountain in Gyirong County in Southern Xizang. It grows close to the ground with leaves spread out over it. Both its leaves and flowers are dark red in color. The lower picture is a veriety found on the same mountain at a lower elevation, 4800 meters a. s. l. Its has large fat root penetrating deep into the soil.

Myricaria prostrata Benth. et Hk. f.

This is a picture of the plant found in the steppe of northern Xizang. The plant is distributed on mountain slopes and plateaus above 4500 meters a. s. l. It grows close to the ground with dense close twigs as a means against the cold and widiness of the alpine environment.

Ptilotrichum canescens (DC.) C. A. Mey

The plant is a semi frutex-like herb covered all over with thick stellate downs and distributed in the Ali (Ngari) and Qiangtang districts in Western and Northern Xizang in grasslands between 4800-5100 meters a. s. l.

Saussurea obovallata (DC.) Edgew.

It is found between 4000—4500 meters a. s. l. Its yellow-green membranous bracts viewed from afar on snowcovered mountain slopes look like a lotus flower grown out of the snow. That is why it gets the name "snow lotus". *Saussurea obovallata* (DC.) Edgew. is also used as a medicine, having cooling, soothing and narcotic effects.

Saussurea gnaphaloides (Royle) Sch.-Bip.

This is a common alpine plant in Xizang. It grows in the crevices of small stones below the snow line, but according to our findings, it has been found in places as high as 5900 meters a. s. l.

Saussurea stella Maxim.

   The purple-red bracts of its flower are densely grouped and at the foot of the flower stalk, its green long narrow leaves radiate out in all directions close to the ground. It looks like a shining star in the alpine green grassland. The above is a picture of it. Used medically, it is said it cures toxic fever and broken bones. The above is a picture of the plant found in alpine meadows and swamp meadows.

Saussurea simponiana Lipsch.

   The plant is found on gravel mountain slopes between 5300-5750 meters a. s. l.

A Sea of Clouds in the Forests

Leontopodium nanum (Hk. f. et. Th) Hand.-Mzt.
The plant is covered all over with thick downs as a protection against the alpine cold and windy climate.

Gagea pauciflora Turcz.

Oreosolen wattii Hk. f.

A kind of alpine plant, it grows with its leaves spread close over the surface of the ground. It has yellow flowers and is found on mountain slopes ranging from 4500 to 5300 meters a. s. l.

Waldheimia glabra (Decne.) Kitam. (syn. *W. tridaclylites* Kar. et Kir.)

An alpine plant of the Xizang plateau and the Central part of Asia, it is found in Xizang on mountains at altitudes from 4900 to 5500 meters.

Alpine Cushion Plants

Some plants can not grow high above the ground under the alpine conditions of severe cold, strong wind and intense solar radiation. They are forced to grow low, close to the ground, with thick branches and twigs intertwined together in the form of a cushion. The spread-out flat form allows them to absorb more heat under sunshine. At night it makes the dissipation of heat less rapid, but gradual, thus reducing greatly the temperature difference between day and night. At the same time the closely knit branches and twigs prevents rapid and excessive evaporation of its water content. Their stream-lined configuration offers minimal resistance to the blow of wind, thus reducing damage from the latter to a minimum. Alpine plants taking this form are called cushion plants. The cushion form is an adaptation to natural environment.

Arenaria musciformis Edgew.
It is found on mountain slopes at altitudes from 4700 to 5400 meters.

Androsace tapete Maxim.

It is found on mountain slopes at altitudes from 4700 to 5300 meters.

Thylacospermum   caespitosum (Camb.) Sch. (syn. *T. rupifragum* (Fenzl) Shrenk.)

It is distributed at altitudes from 5000-5400 meters

Androsace hookeriana Klatt.

It is distributed in Nyalam County, Xizang at 4000 meters a. s. l.

Lichenes

Lichens are a kind of lower plants growing in colored patches on stones, trees, and soil. They are not only adapted to extremely unfavorable ecological conditions, but also secrete a kind of lichen acid that dissolves rocks and wood for absorbing nutrients. This dissolution accelerates the weathering of the rock and turns it into primitive soil. Therefore, lichens are labeled as vanguard plants. Some lichens can be used for the determination of the age of glaciers and others are regarded as indicators of environmental contamination.

Sporastatia testudinea (Ach.) Mass.

The distribution of this plant is concentrated on rocks on the northern side of Mount Qomolangma. It is an Arctalpine element. The above picture is taken at 5500 meters a. s. l.

Lethariella sinensis Vej et Jiang sp. nov.

It belongs to the sub-genus *Chlorea* of Genus *Lethariella*. Now it is discovered in Qando and Riwoqe (Leiwuqi) Counties. It is an alpine species.

Sub-Genus *Chlorea* has six species, four of which are distributed in the Sino-Himalaya and one in the Canary Islands. The present discovery provide a new evidence for the connection between the flora of Xizang and the flora of the ancient Mediterranean region.

The above picture is the lichen grown on a cypress tree in Qandu County, Xizang, at 4300 meters a. s. l.

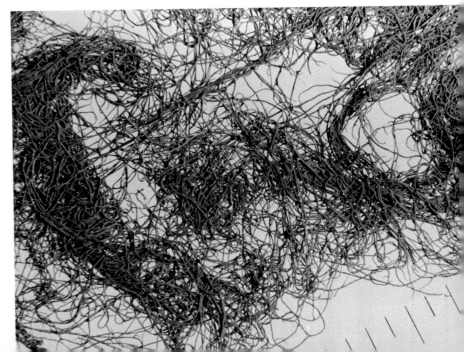

68

Lethariella flexuosa (Nyl.) Vej. et Jiang.
This lichen is an endemic species to the Himalayas. It grows on brushes at 5000 meters a. s. l.

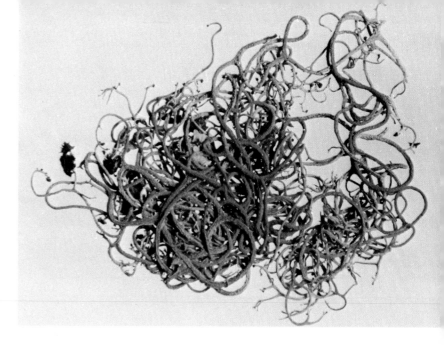

Rhizocarpon tinei (Tornab.) Run.
It is found on rocks on the northern slope of Mount Qomolangma. It is a multirangés element of the holoarctic areal type. It is used for determining the active age of glaciers.

Umbilicaria hypoccinea (Jatta) Llano
It is a mountaine element of the Eastern Asian areal type, a Sino-Himalayan subtype. It is widely distributed in the Wutai Mountain, Shanxi Province, the Qin Ling Mountains, Shaanxi Province, and the Himalayas on rocks above 3200 meters a. s. l.

Alpine Pasture Lands

In high mountains, the lands above the forest line are usually covered by lushy communities of herbs. As the alpine temperature difference between day and night is great, the high temperature during the day is favorable to the synthesis of organic matters, and the low temperature at night reduces the consumption of organic matters. These two factors combine to enhance the accumulation of organic matters. That is why alpine grasses are often very nutritious. They are good fodder for animals. Alpine plant communities often used as pasture grounds are steppe, meadow, desert steppe and swamp meadow.

The above is a picture of a pasture meadow of *Kobresia*.

A pasture of swamp meadow

This is the picture of a swamp meadow pasture on the shore of the Qi-Xiang Cuo Lake in Northern Xizang, at 4600 meters a. s. l.

A pasture of steppe of *Stipa purpures* Griseb.

Pastures of *Stipa purpurea* Griseb. are widely found on mountains and in the plateau in Northern Xizang at altitudes from 4400 to 5200 meters a. s. l. and in Southern Xizang at altitudes from 4800 to 5300 meters.

Stipa purpurea Griseb.

This grass grows at altitudes from 4400 to 5400 meters. It is a community forming plant in alpine steppe and a very good pasture grass.

Kobresia pygmaea C. B. Clarke

This herbaceous plant is the main element of alpine meadows. It is soft, tender, and nutritious. It contains about 15% protein. It is good for grazing and has good adaptability to pasturing. It is a specially good forage for the yak. It is an excellent forage for fattening the local domestic animals.

Carex moorcroftii Falc. ex Boott

It is an alpine plant endemic to the Qinghai-Xizang Plateau. It possesses great adaptability to different ecological environments: it can grow on damp swampy meadows, arid steppe, and sand-covered slopes. It is often used as a pasture grass.

The above is a picture of the plant on the shore of the Zharinanmucus Lake in Northern Xizang, at 4650 meters a. s. l.

Swamp meadows

These are the pictures of two swamp meadows. The upper photo is a swamp meadow of *Kobresia deasyi* C. B. Clarke (syn. *K. pamiroalaica* Ivan.) located in the upper reaches of the Yarlung Zangbo River, at 4500 meters a. s. l., and the lower one is a swamp meadow of *Kobresia littledalei* C. B. Clarke, located in the Qiangtang Plateau in Northern Xizang, at 4600 meters a. s. l.

Blysmus sinocompressus Tang et Wang
   A principal component of swamp meadows on lake shores and sandy river banks, the plant is endemic to China. It is also a pasture grass.
   The above is a picture of the plant grown on the shore of the Nam Co Lake, Xizang, at 4730 meters a. s. l.

Kobresia deasyi C. B. Clarke (ayn. *K. pamiroalaica* Ivan.)
   It is a principal component of swamp meadows in Western Xizang, and also an important pasture grass.

Polygonum viviparum L.

The seeds of the plant germinate before they are shed from the plant. The germinated seeds will grow into new plants after they fall off the mother plant down to the ground. That is why the plant is named viviparum. This kind of germination is the result of long adaptation to adverse ecological environment. Both its seeds and root stock are rich in starch. So it is a good fodder for fattening domestic animals. They may be used as raw material for making wine or serve as a substitute for grain in a famine. The plant is an arcticalpine element and grows in alpine meadows and under brushes.

Urtica hyperborea Jacq. ex Wedd.

The plant grows on desert lands and between stones at altitudes from 3800 to 5000 meters. It is covered all over with bristling downs that cause both pain and swelling when touched. But it can be used as a fodder after boiling and will increase the milk of mother animals. It is an endemic plant to the Qinghai-Xizang Plateau. Used medically, it is said it cures rheumatism and insect bites.

Caragana jubata (Pall.) Poir.

It is a kind of alpine deciduous brush of the family Leguminosae with stings. It is an economic plant. Used internally, it reduces hypertension; Used externally, it cures boils and swellings; its flowers cure headache and cough and its root can cure all kinds of body injuries and rheumatic arthritis. Its fibers are used as material for making ropes and gunny-bags.

Arisaema tortuosum (Wallich.) Schott

Delphinium brunonianum Royle

It is an alpine plant distributed in the central and western parts of Xizang, India, Nepal, and Afghanistan at altitudes ranging from 4500—6000 meters. A medical herb, it is said it can cool blood and detoxify poisoning, cure influenza, itch rash, and snake bite. Its seeds contain 30 % oil that can be used as an industrial oil.

Soroseris gillii (S. Moore) Stebb. [syn. *S. hookeriana* (C. B. Clarke) Stebb.]

Found in alpine meadows at 5500 meters a. s. l. Used as a medical herb, it is said it has the virtue of antipyretics, antidotes, and narcotics, is effective against rheumatism and laryngitis, and heals all sorts of internal physical injuries.

Sophora moorcroftiana (Wall.) Benth. ex Baker
   It is a short deciduous shrub with stings of the family Leguminosae found in the sub-alpine belt. Used medically, it is said it is good for the stomach, and is an antipyretic, an diuretic, and an insecticide.

Ceratostigma minus Stapf.
   It is often found in the bushes of *Sophora moorcroftiana* (Wall.) Benth. ex Baker. It is distributed in Xizang, Sichuan, Yunnan, and the southern part of Gansu Province. Its roots are said to be effective against inflammation, aches, and rheumatism.

Berberis sp.

It is a kind of deciduous shrub grown in the glades of forests. Because it has trident stings under its leafstalks, it got the name "Trident stings". Alkaloid of *Berberis* can be extracted from both its root and bark, namely the so-called extract of *Coptis chinensis* Franch. now widely used as a medicine. The extract is effective against stomatitis, laryngitis, conjunctivitis, acute and chronic enteritis, and dysentry. It is generally antipyrectic and antitoxic.

Anisodus luridus Link et Otto

Its growth requires comparatively fertile soil. Medically, its root has the virtue of stopping convulsion and pain. It is also a cure for stomachache, cholecystitis acute and chronic enterogastritis. Over dose causes palpitation and dilatation of the pupil. It is a source of atropine.

Incarvillea younghusbandii Sprague

In alpine steppe above 4500 meters a. s. l. there are large purple red flowers high above the ground. This is the plant *I. younghusbandii* Sprague, endemic to Xizang. its root and seeds are medicines that are said to be a cure for dizziness, anaemia, general weakness after protracted illness, and lack of milk after birth. It is a general tonic for the body.

Dracocephalum heterophyllum Benth.

With pale yellow labiate flowers, the plant is used as a medical herb. It is said it is a cure for headache, cataract and general internal heat. It is found in alpine steppe or alpine meadows above 4000 meters a. s. l.

Ephedra saxatilis Royle

The plant is a undershrub 20—60 cm high, found in gravel-sand terrace lands 3000—3500 meters a. s. l. in the valley of the middle and lower reaches of the Yarlung Zangbo River. Its leaf has degenerated into membrane. It is drought-resistant. Used medically, it is said it can stop cough and pain, hemorrhage, cure asthma and dropsy. It is also a diuretic and a sudorific. As a sudorific, it cures common cold.

Verbascum thapsas L.

A perennial plant of the family Scrophulariaceae it is covered all over with thick yellow downs. It is found on mountain slopes, river beaches and terraced lands at 3000—3500 meters a. s. l. Medically, it is said it cures inflammation, stops hemorrhage, remove internal heat, and neutralize toxin. It can also cure pneumonia, injury of the joints, and skin ulcers.

Rheum nobile Hk. f. et Th.

    An alpine perennial herb it grows at altitudes ranging from 4200—4600 meters, Medically it is said it cures constipation, dysentery, dropsy, and certain types of disease caused by dampness, dissolyes all sorts of silting or accumulation such as indigestion, and revive circulation of blood, Used externally, it is effective against boils, malignant boils, scalds and burns.

Arisaema flavum Schoot

The plant is found on mountain slopes, farmland edges, and roadsides in the subalpine steppe belt ranging from 3500 to 4300 meters a. s. l. Its bulb in poisonous. It can be used as a pesticide, Used medically, it is said it can cure chronic tracheitis, bronchiectasis, tetanus, and epilepsy.

Astragalus tibetanus Benth. et Bge.

It grows on dry southern slopes in subalpine steppe belt ranging from 3500—4500 meters a. s. l. Its root is said to be effective for curing indigestion, dropsy, and prostration.

Fritillaria cirrhosa Don

It grows on mountain slopes ranging from 3000—4000 meters a. s. l. Its bulb tastes bitter-sweet and is used as a medicine for curing cough, clearing phlegm, eliminating internal heat, and nourishing the lungs.

Pterocephalus hookeri (Clarke) Höeck
Its has clusters of white flowers in the shape of spheroid. Its root is said to have the virtue of eliminating internal heat and rheumatism; neutralizing toxin, and stopping pain and fever of common colds. It is a kind of concommitant plant of alpine steppe.

Rheum palmatus L.
Its root contains aromatic oil and is used as a medicine for curing pyretic toxin, breaking up various forms of congestion, dissolving extravasated blood, curing constipation caused by real internal heat, dysentery, dropsy, bloodshot eyes, headache, and amenorrhoea. Used externally, it can cure carbuncles, malignant boils and scalds and burns.

Saussurea tridactyla Sch.-Bip.

This plant has a special form to adapt itself to the alpine environment of growth: in addition to the lotus flower configuration, it is also covered all over with thick downs. So it is better protected against the alpine cold, as is shown in the upper picture. The picture below shows it often grow on gravel slopes near the snow line. It is as beautiful as a lotus flower on a snowcovered plain. It is also a medical herb. It is said it has the virtue of enhancing man's virility, enriching the blood, and warming the womb. It cures cough, expectoration, impotence, sexual debility, weakness of the pancreas, lumbago, irregularity of menstruation, and menorrhagia and metrorrhagia.

Cordyceps sinensis(Berk.) Sacc.

It is formed by the parasitism of the fungus *Cordyceps sinensis* (Berk.) Sacc. on the larva of *Hepialus virescens*. The larva hibernates underground in winter and the spore of the fungus enters the body of the larva and feeds on it and causes its death. At the end of spring and the beginning summer of the next year, the spore of the fungus will grow out of the head of the larva and come out of the ground like a little grass. That is why it got its Chinese name "Dongchong Xiacao", meaning winter larva, summer grass. It is generally found in alpine meadows over 4000 meters a. s. l. It is a well-known tonic. It tastes bitter sweet and is mild in its virtue. It strengthens the lungs and enhance the power of virility. The picture at the upper left corner is the plant—*Cordyceps sinensis* (Berk.) Sacc.

Gastrodia elata Bl.

It is a saprophyte of the family Orchidaceae, Owing to the existence of parasite fungus, its rhizome is enlarged like a stem tuber, is fleshy and tastes sweet. It is used as a medicine for curing headache, giddiness, blurred vision, infantile convulsions, and trigeminal neuralgia.

Hypericum bellum L.

It is a kind of shrub grown in glades of forest. Its fruit has the virtue of eliminating internal heat, neutralizing toxicity, killing worms, and stopping itching. It cures acute and chronic hepatitis, common cold, dysentery, stomatitis, roundworm and dermatitis.

Panax pseudoginseng Wall.

It grows in dark damp places in thick forest. Its meaty rhizome is very effective for healing internal and external injuries and stopping haematemesis, metrorrhagia, and bleeding. It also dissolves extravasated blood, resolves swells, stops aches and pains. It is a potent general tonic. It is a well-known very effective styptic.

88

Stellera chamaejasme L.

It is a common herb in alpine steppe. Its flowers, red outside and white inside, grow in tufts in the shape of spheroids and are beautiful. Its root is poisonous and is said medically effective for killing parasite worms reducing inflammation, stopping pain or ache and curing skin diseases. Its fibers are excellent for paper making. The upper picture is a close shot of it.

Primula sikkimensis Hk.

It is found on watersides. Its beautiful yellow bell flowers hang down gracefully and make it look like a blushing young maid standing on the bank of a river. It is also a medical herb, effective for curing acute gastritis and dysentery. It is also a haemostatic.

Lagotis integra W. W. Sm.

It grows on stony slopes ranging from 4000 to 5000 meters a. s. l. It is a medical herb effective for dispelling internal heat, neutralizing toxin and lowering hypertension. It is used for treatment of acute and chronic hepatitis. It can be used as a substitute for *Coptis chinensis* French, the rhizome of Chinese goldthread. So it got the name "Xizang Coptis".

Rhodiola fastigiata (Hk. et Th.) Fu
It is plant always growing with alpine bush. It is a medical herb healing all sorts of injuries. The above is a picture of the plant found at 4100 meters a. s. l.

Anemone rivularis Buch. -Ham.
It grows under alpine bush at 4000—4500 meters a. s. l. Its root has the virtue of warming the stomach, stopping emesis, and killing worms.

Bergenia purpurascens (Hk. f. et Th.) Engl.
The whole plant is used as a medicine for stopping various kinds of hemorrhage, curing giddiness, and general physical feebleness.

Meconopsis horridula Hk. et Th.

The entire plant is used as a medicine. It is used for treatment of carbuncle, hepatitis, hypertension and various sorts of pains and aches.

To adapt to the growing cold and dryness of increasing elevation, the plant becomes shorter and shorter as the altitude goes up. The left picture shows the plant near the forest line at 3950 meters a. s. l. The lower right picture shows the plant at 4500 meters a. s. l.

Lamiophlomis rotata (Benth.) Kudo

It is a medical plant curing pain and ache of bones and muscles, healing all sorts of injuries, resolving swells and reviving the circulation of blood. It has purple flowers that grow in layers verticillately around a quadrangular stem, that makes the plant like a purple precious pagoda towering above the green grass out of which it grows.

94

The Swamp of *Hippuris vulgaris* L.

A swamp plant, it is found on the brinks of pools and lakes. It tastes bitter and possesses the virtue of supperssing coughing, soothing the liver and cooling the blood. The above picture was taken at 4500 meters a. s. l.

About the *Rhododendron* and its flowers.

There are 850 species and varieties of the Genus *Rhododendron* in the whole world. In China alone there are already over 460 species identified. They are distributed in damp mountain areas in the southeastern part of Xizang, and the western part of Yunnan. The whole region has earned the title "The kingdom of Rhododendron", The flowers of all the species and varieties of the Rhododendron are gorgeously rich in color and form. Roughly speaking, there are scarlet, pure white, orange yellow and indigo purple flowers, but between these there are all shades of color beyond imagination. As the rhododendron bears such beautiful flowers, it is extensively cultivated in gardens and homes. The flowers of some species of it look so much like the rose and as fragrant that they are called "alpine roses". Like the genuine rose, aromatic oil is also distilled from them which is used in the treatment of certain diseases.

Rhododendron neriiflorum Franch.

Rhododendron principis Bur. et Franch.

Rhododendron baileyi Balf. f.
A kind of evergreen shrub with red flowers, it is found in the southern part of Xizang at 3100—4300 meters a. s. l. It is also found in Bhutan.

Rhododendron setosum D. Don
With purplish rosy color, it is distributed in the southern side of the middle section of the Himalayas at altitudes ranging from 3500—4800. It is also found in Sikkim.

Rhododendron companulatum D. Don aff.
It is an evergreen shrub (picture below) with white or rosy flowers (picture above). It is distributed in the southern part of Xizang, also in Sikkim, Bhutan, and Nepal.

99

Rhododendron cephalanthum Franch.
    A short evergreen brush bearing white flowers, it is distributed in the southeastern part of Xizang, the western part of Yunnan, the southwestern part of Sichuan at altitudes ranging from 3000 to 4400 meters (See the two upper and lower pictures).

Rhododendron arboreum Sm.

An evergreen tree, it has two varieties of white and rosy flowers respectively. It is found on the southern side of the middle and the eastern sections of the Himalayas at altitudes ranging from 2300-4200 meters. It is also found in India, Bhutan and Sri Lanka.

Rhododendron nivale Hk. f.

It bears bright purple, light red, bright purplish blue, and clove purple flowers and is distributed in the southeastern part of Xizang and Sikkim from 4000 to 4800 meters a. s. l. An evergreen short shrub, it is an important component of alpine bush.

101

Rhododendron lepidotum Wall. (syn. *Rh. elaeagnoides* Hk.)

An evergreen shrub with purple, deep red, rosy, pale yellow, light green, and white flowers, it is distributed in the southern and the southeastern parts of Xizang from 1700 to 4200 meters a. s. l. also in Nepal, Bhutan and Sikkim. It is a poly-typical species of plants.

Rhododendron wightii Hk. f.
　　A evergreen shrub with yellow and white flowers. it is distributed in the center section of the Himalayas from 3900 to 4300 meters a. s. l.

Rhododendron campylocarpum Hk. f.
　　It got its name from the crescent shape of its fruits. An evergreen shrub with yellow flowers, it is distributed on slopes at 3200-3500 meters a. s. l. in the sonthern part of Xizang, also in Nepal and Sikkim.

Cotoneaster divaricatus Rehd. et Wils.
   A deciduous shrub with rosy flowers, it is distributed in eastern
Xizang on mountain slopes from 3300—3900 meters a. s. l.

Cotoneaster microphyllus Wall.
   An evergreen short shrub with branches loaded heavily with
small red fruits, it grows on mountain slopes at altitudes from 2500 to
4100 meters.

Meconopsis paniculata (D. Don.) Prain

Phyllophyton decolorans (Hemsl.) Kudo.
　　Covered all over with soft white downs and its leaves put in order imbricated closely on the stem, it is a creeper on stony mountain slopes above 4500 meters a. s. l.

105

Dicranostigma lactucoides Hk. f. et Th.

Geranium refractum Edgew. ex Hk. f.

An annual herb with white purple flowers, it has creeping stem. The whole plant can be used medically for treatment of rheumatic arthritis, muscular numbness, and external and internal injuries. It has the virtue of strengthening the bone and muscle and clearing away rheumatism.

Cassiope fastigiata D. Don.

    With branches wrapped thickly in little bell-like white flowers that will swing about in the slightest of wind, it is an alpine evergreen bush that grows in damp and cold places in the mountains.

Rhodiola ratundata (Hemsl.) S. H. Fu.

Rhodiola quadrifida (Pall.) Fisch. et Mey.

It shows its tough vitality by growing in the snowcovered grounds at 5200 meters a. s. l. in the mountains. Its fat, juicy, succulent leaves protect it from death of cold dehydration.

Androsace robusta (R. Knuth.) Hand. -Mzt.

Meconopsis pinnatifolia C. Y. Wu et H. Chuan sp nov.

Meconopsis raceorosa Maxim.

109

Anaphalis nepalensis (Spreng.) Hand.-Mzt. var. *monocephala* (DC.) Hand.-Mzt. aff.

It looks like chrysanthemum, but is not it. It grows in alpine meadows 4500 meters a. s. l. The petal-like white things on it are neither flowers, nor petals but are some transparent bracts.

Chesneya purpurea P. C. Li
It is found in Lakang, Xizang, at 4300 meters a. s. l.

Iris potaninii Maxim and its variety—*I. potaninii* Maxim. var. *ionantha* Y.T. Zhao.

It is a common perennial herb in the steppe at 3800—5100 meters a. s. l. It has yellow and purple flowers. The upper picture is the plant with yellow flowers found on stony mountain slopes at the Nyanya Xiongla Pass, 5100 meters a. s. l. Nyalam County, Southern Xizang. The lower picture is a variety of the *Iris potaninii* Maxim. var. *ionantha* Y. T. Zhao, found growing in *Stipa purpurea* Griseb. steppe at 4800 meters a. s. l. in Northern Xizang.

Iris sp.

    It is found in Damxung County of the Nyainqêntanglha Range at 4400 meters a. s. l.

Iris nepalensis D. Don.

    It is found in Dinggye County, on mountain slopes 3100 meters a. s. l.

Dasiphora fruticosa (L.) Rydb.

Dasiphora fruticosa (L.) Rydb. var. *pumila* Yü et Li

It is a variety of *Dasiphora fruticosa* which is a deciduous shrub. Owing to constant adaptation to alpine cold and dry conditions through long ages, it has become a shorter shrub. Its distribution is found to reach 5400 meters a. s. l., but the distribution of the typical species is limited to below 4800 meters a. s. l.

Pedicularis anas Maxim. var. *ti-betica* Bonati

Primula denticulata Smith
It is found in Cuona County, Xizang, at 4000 meters a. s. l.

Anemone smithiana Lauener et Panigrahi
It is found in Nyalam County, Xizang, at 4300
meters a. s. l.

Gueldenstaedtia himalaica Baker
It is found in Nyalam County at 4300
meters a. s. l.

Rosa sweginzowii Konchne

Rosa webbiana Royle
It has large rosy flowers and is cultivated as a garden flower for decoration. It is distributed in Central Asia, the western section of the Himalayas and Xizang. The above picture is the plant with its flowers found in Burang County, Xizang.

Aster asteroides (DC.) O. Kutze aff.

Gentiana sp.

It is found in the alpine meadows at Duoxiongla Mountain Pass at 4200 meters a. s. l. in the eastern section of the Himalayas.

Polygonum affine D. Don
It is found in Cuona County Xizang, at 4300 meters a. s. l.

Nomocharis meleagrina Franch.
It is found in Nyalam County, Xizang at 4050 meters a. s. l.

Lilium nepalense D. Don.
   It is found on the southern slopes of the Himalayas.

Potentilla biflora Willd. ex Schlenchp.
   It is found on the northern slope at
5150 meters a. s. l. of mount Langmalangyi
in the Ali Prefecture.

Meconopsis harleyana Taylor aff.

A Waterfall

Meconopsis florindae Kingdon-Ward

Cirsium bolocephalum Petrak.

It is found at the Duoxiongla Mountain Pass at 4100 meters a. s. l. in the eastern section of the Himalayas in Motuo County, Xizang.

Thalictrum chelidonii DC.
It is found in Gyirong, Xizang, at 3200 meters a. s. l.

Iris goniocarpa Baker

Callistephus chinensis (L.) Nees.

It is often seen in Lhasa, Rigaze, etc. at 3600—3900 meters a. s. l. It is also cultivated as a garden flowering plant.

Dahlia pinnata Car.

It is a flowering plant endemic to America, but is found in Lhasa, 3700 meters a. s. l. as a cultivaled garden flowering plant.

Fuchsia hybrida Voss.

    It is a flowering plant endemic to South America, but is found in a temple in
Zhashi Lunbu, Rigazê is 3900 meters a. s. l.

Contents